Laboratory Fume Hoods Explained – 2026

by

Chip Albright

A Fume Hood Certified Publication

Fume Hood Certified, LLC
PO Box 71477
Phoenix, AZ 85050 USA

www.fumehoodcertified.com

experts@fumehoodcertified.com

+1 623-662-0907

Disclaimer

This publication is designed to provide accurate and authoritative information with regard to the subject matter covered. It is sold with the understanding that the contents do not render professional advice. If you require professional assistance or expert advice, you should seek the services of a licensed professional.

Any company names, logos, or images used that are trademarked or copyrighted by others are acknowledged to be the property of the trademark or copyright owner.

This book was created with assistance from artificial intelligence resources. The content has been reviewed and edited by a human. For more information on the extent and nature of AI usage please contact Chip Albright.

experts@fumehoodcertified.com

Paperback: ISBN 978-1-7357110-9-6

First paperback edition December 2025.

Printed in the USA

Table of Contents

Introduction

Five years ago, when I completed the first edition of Laboratory Fume Hoods Explained, I thought I was closing another chapter of my career. What I didn't realize was that I was only beginning a much larger story, one that would lead to founding a new company, developing new test methods, and embracing a mission that had become personal: **making labs safer, one fume hood at a time.**

Back then, most people saw a fume hood as just a box with a fan. It was a protective fixture, important, but static and unremarkable. It didn't breathe, adapt, or teach. That's changed. In the years since, I've watched laboratories shift their thinking, not just about fume hoods, but about containment itself. We've moved from measuring airflow to visualizing it, from trusting numbers to trusting what we can see.

Let's be honest, safety sounds boring. But the real goal isn't safety for safety's sake; it's accident prevention. The conversation has evolved:

Face velocity ≠ Containment
Compliance ≠ Safety

This 2026 edition is not just an update; it's a leap forward. It reflects how our understanding has grown and how the industry has changed. Inside, you'll find new chapters on

That may sound like a small change, but it's enormous. Face velocity tells you how fast air moves into the hood. Containment tells you whether that air is doing its job, keeping contaminants inside and people outside safe. For decades, we relied on stickers and numbers. But those numbers often hid the truth.

Story from the Field:
It was a Tuesday morning in a university teaching lab. A technician smelled something odd even though the hood's certification sticker was current and the monitor read 100 fpm. Moments later, we ran a Tri-Color visualization test. The air rolled out of the hood like a slow-motion wave. A cross-draft from a nearby door had turned the hood from protector to polluter. The sticker said PASS. The air said otherwise.

Safety can't be standardized into a one size fits all concept. Seeing what were thought to be safely performing hoods fail changed how I teach containment. You can't measure safety with a number; you have to see it. Visualization tools like Tri-Color give us that sight. For the first time, people can see what the air is doing, and more importantly, when it stops doing what it's supposed to do.

Containment is about stability and relationships. When the building system and the user are in tune, the hood performs like a conductor in an orchestra; quietly, precisely, invisibly.

But when balance is lost, chaos takes over. Air becomes noisy, unpredictable, and unsafe.

In a well-balanced system, the hood's job is to create a directional flow, a protective barrier between the user and the hazard. But that barrier is fragile. It depends on pressure differentials, smooth air entry, and user awareness. Even small disturbances, someone walking by, a door opening, can disrupt containment.

When we start to see the hood as part of a breathing system, safety culture begins to change. It's no longer about compliance, it's about understanding. Users begin to notice how their movements, their posture, even the height of the sash influence performance. They become part of the containment process.

Containment isn't something a fume hood does for you. It's something you and the hood do together.

The laboratories that understand this are the safest ones I visit. They don't rely on stickers. They rely on awareness. They visualize airflow, test frequently, and train users not just to operate hoods, but to understand them.

Every safe laboratory breathes in rhythm. The fume hood is the lungs. When everything is in tune, room balance, exhaust flow, user habits, the system works silently, effortlessly. But when something falls out of rhythm, the lab itself starts to struggle to breathe.

Containment cannot be assumed. It must be demonstrated,and it must be demonstrated often. That's what this new generation of visualization technology has taught us. If you can see it, you can fix it. If you can't, you're working on faith.

And faith, hope and luck in the lab, are dangerous things to depend on.

Chapter 2

Fundamentals of Airflow and Containment

"Air is invisible, but its movement tells the truth."

When you stand in front of a fume hood, what you're really looking at is air in motion. You can't see it, but it's there, flowing over your shoulders, sliding over your hands, curling around beakers, and swirling at the sash. Air is the most important safety material in the laboratory, and yet it's the one we understand the least.

Over the past five years, our understanding of airflow has changed dramatically. We've learned that containment is not about how fast air moves, it's about how it behaves. The days of believing that 100 feet per minute meant safety are over. A stable, predictable airflow at 60 fpm can be far safer than turbulent air at 100.

When I first began teaching airflow, I used smoke. It worked, sort of. The problem was that smoke fades, disperses, and lies. You'd see a momentary plume, but it wouldn't tell the whole story. Then we standardized laser visualization technology and for the first time, we could see airflow in color, continuously, without distortion. That changed everything.

Story from the Field:
A few years ago, I was working with a pharmaceutical research group that was having intermittent odor problems. Every test said the hoods were fine. Face velocity readings looked textbook-perfect. Then we brought in the Tri-Color Visualizer. Within seconds, the truth appeared. The airflow rolled outward from the center of the sash, like a slow leak in a dam. The airspeed was there, but the containment wasn't. It was turbulence, the invisible enemy, that caused the leak.

Visualization made what was invisible undeniable. It also changed how people thought. Once they saw it, they couldn't go back to trusting numbers alone.

The Nature of Airflow
Air doesn't move in straight lines. It swirls, bends, and reacts to everything around it, temperature changes, people walking by, even the heat from your hands. Inside a fume

hood, airflow is shaped by geometry, baffle design, and the subtle dance between supply and exhaust. If one part of that balance is off, the entire airflow pattern changes.

To understand containment, we have to accept that airflow is alive. It responds, it reacts, it adapts.

Analogy:
Think of airflow like a river. When it's calm and steady, it carries everything downstream. But when rocks or debris get in the way, eddies and whirlpools form, and suddenly water moves where it shouldn't. The same happens inside a fume hood, turbulence becomes the enemy of control.

Dynamic Containment vs. Static Velocity
Static testing measures speed; dynamic testing measures behavior. The difference between the two is the difference between a heartbeat charted on paper and a living breathing person. Visualization gives us that heartbeat in motion. It shows how air responds when a sash moves or when an arm reaches in. It lets us see the system under real working conditions.

That's why visualization is now considered best practice worldwide. Numbers can tell you that something failed, but visualization shows you why it failed.

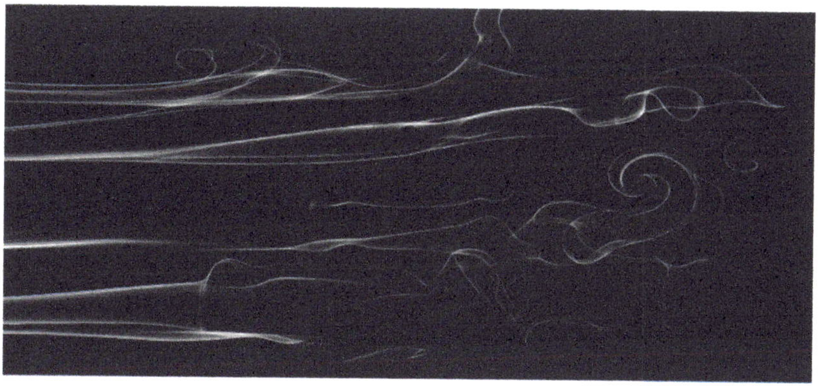

Turbulence and Wake Zones

Every hood contains invisible pockets of instability, what I call wake zones. These are the low-pressure regions where air can swirl and roll back toward the user. They often form around obstructions, behind equipment, or at the edges of baffles. In the past, we couldn't identify them even with complex tracer gas systems. Now, with lasers, we can see them instantly.

In one demonstration, a Tri-Color beam illuminated a swirling pocket just above a hotplate. Every few seconds, a green shimmer escaped past the sash. The crowd of safety officers gasped. They weren't watching data, they were watching a hazard in real time.

Key Insight:

Turbulence is invisible until you see it. Once you do, you can control it.

Visualization as a Teaching Tool
One of the greatest benefits of visualization isn't just in testing, it's in teaching. When students see airflow respond to their movements, they instantly understand containment. You can lecture for hours, but a ten-second visualization teaches for life. I've watched young researchers adjust their habits on the spot, lowering the sash, slowing their motions, because they saw the air react.

That's the power of making the invisible visible.

The Laboratory Symphony
I often compare a laboratory to a symphony. Every element, fans, ducts, diffusers, hoods, and people, plays an instrument. When one goes out of tune, the harmony collapses. Visualization lets us see where the music falters. It helps us retune the system before anyone gets hurt.

Key Insight:
Airflow may be invisible, but containment failures are not. Seeing the movement of air allows us to finally understand containment as it truly is, a living, dynamic process.

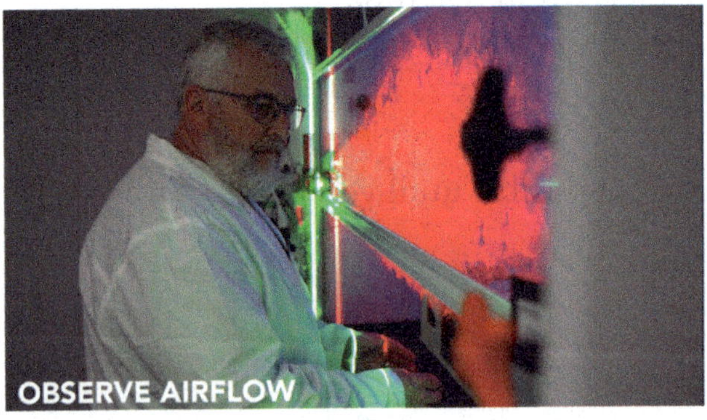

Chapter 3

Evolution of Fume Hood Design

"Design is no longer about shape; it's about how air behaves within it."

When I began in this industry, a fume hood was a box with an exhaust fan. It wasn't elegant, it wasn't intelligent, and no one thought of it as an engineered system. It was simply a way to pull contaminated air away from a worker's face. But over time, I learned that even a box has a personality, and that personality lives in its airflow.

In the years since the first edition of this book, the design of fume hoods has evolved faster than I ever imagined. We've moved from intuition to data, from guessing to proving. Today, every curve, lip, and angle inside a hood is there for a reason, and every reason is supported by science.

The Birth of Intelligent Design

Modern fume hoods are not passive boxes; they are precision containment systems. The introduction of Computational Fluid Dynamics (CFD) was a turning point. For the first time, engineers could see the invisible. They could visualize how air moved through a hood long before metal was cut. I still remember sitting in a conference room, watching a simulation of airflow inside a virtual hood. The blue streamlines danced like smoke through an orchestra of geometry. Every small change, every radius, every corner, made a visible difference. But CFD has limits. The granularity of the model causes minor turbulence to be filtered out. While the airflow shown might be directionally correct, many of the small details are lost. And it is those small details that make the difference between a good hood and a great hood. That is where laser visualization took us to the next level of understanding.

Key Insight:
Design is not guesswork anymore. It's evidence-based airflow engineering.

The Science of Airfoils and Baffles
The baffle system is the heart of airflow control. It determines how air enters, travels, and exits the hood. Today, engineers now fine-tune airflow paths to minimize

turbulence and eliminate recirculation zones. You want slot velocity to be even left to right and bottom to top. A well-designed baffle system will give an excellent face velocity profile.

After the baffle system, the upper bypass has the most impact on performance. We can use that to manipulate the air to wash the back of the sash, to minimize the vortex and to create a two-zone system.

Then we move on to the front airfoil, the curved edge along the work surface, has evolved. That subtle curve isn't cosmetic. It guides airflow smoothly into the hood, preventing eddies that can spill contaminants back toward the user.

The design shifts to the sash handle and the side posts. The closer to laminar we can get the airflow, the more robust containment will be. For decades we learned to improve designs most by trial and error, but today the design evolution is driven by science.

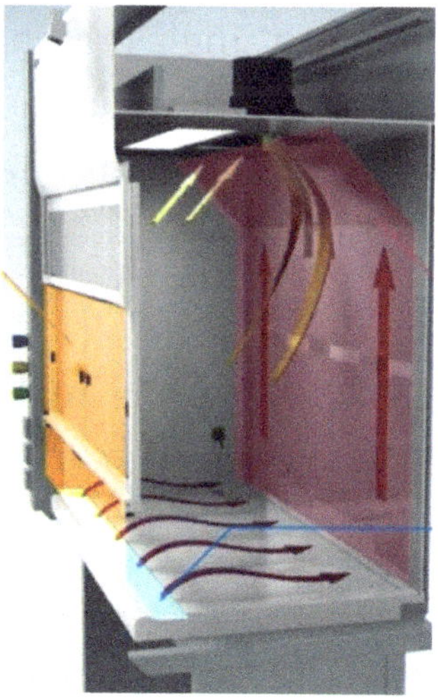

Energy Efficiency and Low-Flow Design

If the early 2000s were about safety, the 2020s are about sustainability. Laboratories consume enormous energy through ventilation. Each hood can exhaust up to 1,000 cubic feet of conditioned air every minute. Reducing that flow safely has become both an environmental and financial imperative.

Low-flow and variable air volume (VAV) hoods usually achieve containment at as little as 60 fpm. A few years ago, that would have been unthinkable. The difference is in the design. These hoods aren't just moving less air, they're moving smarter air. They rely on precisely balanced geometry, real-time feedback systems, and validation through visualization. People are realizing that the room conditions are as important as the hood design.

Story from the Field:

A major pharmaceutical company decided to reduce its fume hood face velocity from 100 fpm to 60 fpm across an entire facility. Everyone was nervous, until we tested it. With Tri-Color visualization, we watched the air move, steady and contained. Not a single breach. In the first year, the facility saved hundreds of thousands of dollars in energy costs. Safety improved, confidence grew, and everyone saw the proof with their own eyes.

Safety vs Energy

Back to the "One Size Fits All" problem. Every room and every lab is different. A university took an entirely different approach. First, they used the ASHRAE LDSL (Lab Design Safety Level) to determine the level of risk available. Then they went lab by lab and did extensive Tri-Color testing. In each room the hoods were slowed down until there was containment failure and then turned the flow back up depending on risk. Lab A might be set to 82 fpm and lab B to 78 fpm and lab C to 90 fpm. Much like controlling blood pressure, each patient may have a different dosage. Now they have safe labs and are saving energy.

The Era of Smart Hoods

The next leap in fume hood design came with embedded intelligence. Smart hoods now monitor themselves, tracking airflow, sash position, and containment stability. Some can even predict when a performance issue is happening. Imagine a hood that can tell you it's drifting out of balance before it fails. That's no longer science fiction.

Sunway Technologies, working in partnership with global experts, have developed what I call the next generation of containment: the Smart Hood. It integrates Tri-Color visualization directly into the hood body and can validate containment with a push of the button.

Performance is monitored using predictive containment that changes the logo color. When airflow is optimal, the logo glows green. When it's compromised, it turns first yellow and then red. Simple and visual than can be seen from across the lab.

Key Insight:
The Smart Hood doesn't wait for you to test it, it tests itself.

Materials and Construction Innovations
40 years ago we thought of hoods more like furniture than containment devices. But today we understand that fume hoods are more like pumps. Given air is a fluid, how it moves the air through it is more pump-like and definitely not furniture.

The materials used in hood construction have also changed. Composites, polypropylene, and reinforced laminates now provide strength and chemical resistance. These materials eliminate corrosion and have excellent flame spread characteristics.

These advancements have reduced noise, improved ergonomics, and extended service life. But the most important evolution isn't in the materials, it's in the mindset. Designers now approach containment as a holistic system. Every component, from sash handle to exhaust plenum, serves the same purpose: keeping users safe.

The Human Element in Design
Technology has given us incredible tools, but the human factor still drives the outcome. A beautifully designed hood can't save a careless user. That's why modern design increasingly focuses on communication, visual cues, audible alerts, and intuitive feedback that teach and reinforce safe behavior.

I often say the best-designed hood is one that trains the user every time it's used. When it glows green, you know you're safe. When it flashes red, you know it's time to pause. That's not just design, that's teaching through technology.

Key Insight:
The future of fume hood design is human-centered. The hood must not only protect, it must communicate.

Looking back, it's hard to believe how far we've come. From noisy, clunky exhaust boxes to intelligent containment systems that think, respond, and teach, have transformed into something extraordinary. It's not just part of the lab anymore. It's part of the learning process.

When I stand in front of a modern hood and see that steady green glow, I'm reminded of how many years it took us to get here. Containment isn't an accident of design; it's the art of airflow made visible.

Chapter 4

Laboratory Ventilation Systems

"A fume hood is only as strong as the air that feeds it."

If the fume hood is the lungs of the laboratory, then the ventilation system is its heartbeat. Air must move in harmony, steady, balanced, and rhythmic. When one part of that system falters, the entire lab feels it. A hood can only perform as well as the air that feeds it.

I've seen labs where everything looked perfect, the hoods tested fine, the monitor was steady, but containment still failed. The culprit? The building itself. Ventilation balance had shifted after a renovation, and the hoods were starving for make-up air. The room airflow shifted when the building shifted from air condition to heat mode. The result was subtle but dangerous: containment that came and went like a tide.

Key Insight:
Containment isn't just a hood issue. It's a building issue. We say we are testing the hood, but the hood hasn't changed. It is the room conditions that changed.

The Laboratory as a Breathing System
A laboratory is alive. It inhales supply air and exhales exhaust. When that rhythm is stable, everything inside

works as intended. When it's disrupted, even the best-engineered fume hood can fail. I call this the 'breathing balance' of the lab.

Every opening, every diffuser, every exhaust fan, supply air fan and VAV controller plays a part in this delicate dance. If one component moves out of sync, if a door opens suddenly, if a diffuser blows across a hood face, room conditions change and become unbalanced. This instability effects containment.

Principles of Ventilation Balance

Good ventilation starts with direction. Air should always flow from clean areas to less clean areas, carrying contaminants away from people and toward the exhaust. That's the heart of containment: direction, not speed. Too often, people think more air equals more safety. In reality, too much airflow can cause chaos, drafts, turbulence, and noise.

Modern laboratories use smart control systems to maintain balance automatically. Variable Air Volume (VAV) controls adjust airflow as sash positions change, keeping the pressure and containment stable. These systems are the silent guardians of lab safety. But VAV adds a level of complexity and without regular calibration and maintenance, VAV causes more problems than it solves.

Story from the Field:

In one facility, technicians complained about inconsistent

hood performance. We discovered high velocity supply diffusers were aimed directly at the sash openings. The hoods weren't broken; they were fighting the room. By redirecting the supply air and recalibrating the balance, the containment became steady. It wasn't a hood fix, it was a system fix.

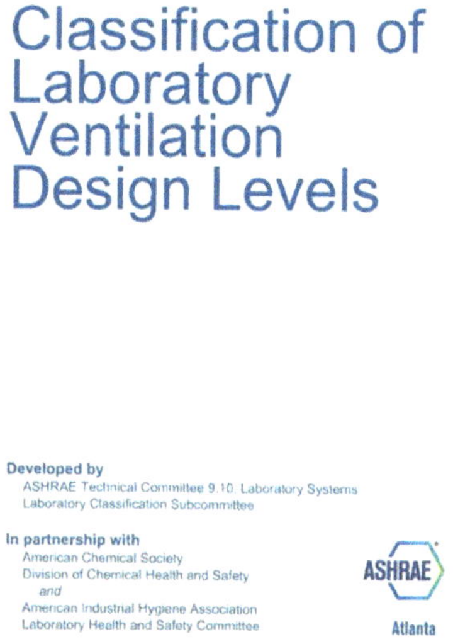
Classification of Laboratory Ventilation Design Levels

Developed by
ASHRAE Technical Committee 9.10, Laboratory Systems
Laboratory Classification Subcommittee

In partnership with
American Chemical Society
Division of Chemical Health and Safety
and
American Industrial Hygiene Association
Laboratory Health and Safety Committee

ASHRAE
Atlanta

ASHRAE's Laboratory Ventilation Design Levels (LVDL)
One of the most important shifts in the past five years has been the adoption of risk-based ventilation design. ASHRAE introduced the Laboratory Ventilation Design Levels, LVDLs, as a framework to match ventilation intensity to the level of hazard.

For example:
- LVDL-0: Low hazard, educational spaces
- LVDL-2: Research and analytical labs
- LVDL-4: High hazard or unknown materials

This approach ensures that air isn't wasted where it's not needed and that high-risk areas receive the ventilation they deserve. It's a smarter, safer way to think about design.

Key Insight:
Ventilation should be proportional to risk, not tradition.

LVDL	Quantities of hazardous materials	Potential for Airborne Generation	Hazard Severity	Control Measures	Construction and Operating Costs	Types of laboratories / application
0	Negligible	Negligible	Negligible	Dilution Ventilation	Low	Computer and Instrumentation laboratories, secondary school teaching laboratories
1	Small	Negligible	Low	Dilution ventilation and local exhaust ventilation	Moderate	Secondary school teaching laboratories, shop areas with point sources of hazardous chemicals, biology laboratories without hazardous chemicals
2	Moderate	Low	Low to Moderate	Pull push dilution ventilation, local exhaust ventilation, fume hoods	High	Academic teaching laboratories, undergraduate research laboratories
3	Large	High	High	Pull push dilution ventilation, local exhaust ventilation, fume hoods, enclosure devices	Very High	Inorganic / organic synthesis laboratories, typical research laboratories
4	Very large	Very High	Extreme	Pull push dilution ventilation, local exhaust ventilation, fume hoods, enclosure devices	Extreme	Inorganic / organic synthesis laboratories, special high-hazard research laboratories, polymer synthesis laboratory, chemical development laboratories

Intelligent and Adaptive Ventilation Systems

The newest ventilation systems are learning systems. They sense occupancy, sash movement, and even temperature to adjust airflow dynamically. They communicate with fume hoods, sharing data to ensure stability. Some systems can even predict when balance will drift and correct it before anyone notices.

I once toured a facility where the control software displayed a real-time map of airflow across the entire lab. Each hood was color-coded: green for stable, yellow for warning, red for imbalance. The engineers could see containment across an entire floor at a glance. It was like watching a living organism maintain its own equilibrium.

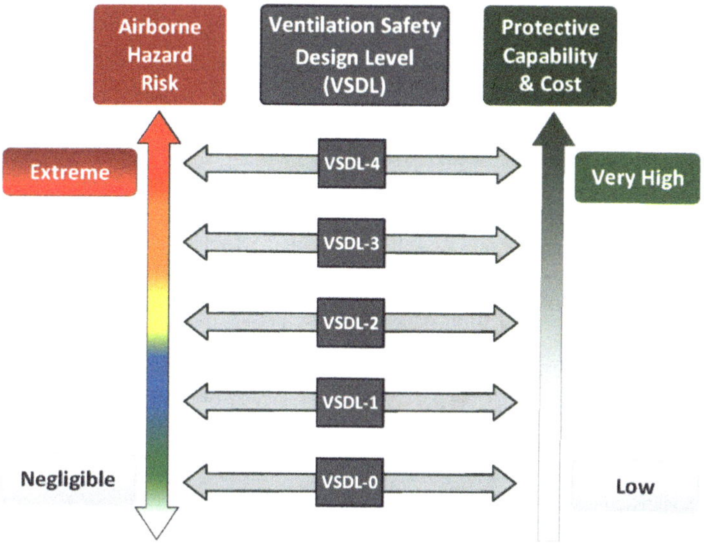

Achieving System Harmony

Containment isn't about power, it's about harmony. Every diffuser, fan, and sash has to work together like instruments in a symphony. When one plays too loud or too soft, the whole performance suffers.

Visualization has become an invaluable tuning tool. Tri-Color tests now help verify system balance visually. Instead of relying on spreadsheets and pressure readings alone, engineers can literally see airflow direction and stability during commissioning.

Key Insight:
The best laboratories are not the ones with the most air, they're the ones where every cubic foot moves with purpose.

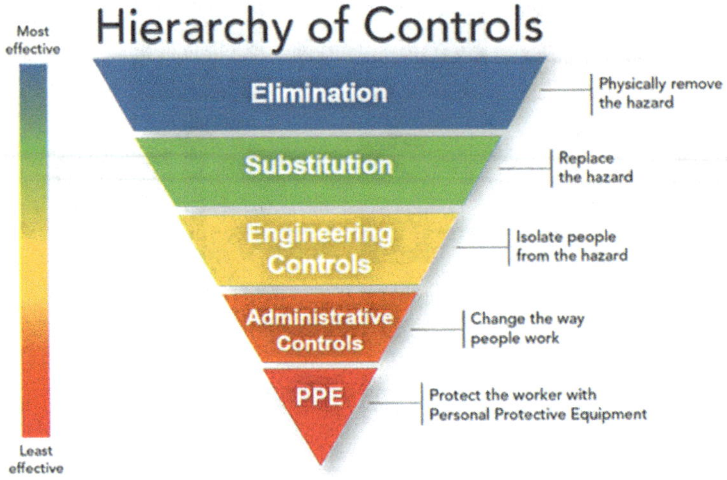

Commissioning and Validation
When a new lab is built or modified, commissioning is the proof of performance. It's not just a checklist, it's a conversation between the building and the hoods. Testing must confirm that airflow is balanced, that negative pressure is maintained, and that containment holds under real use.

Visualization has revolutionized commissioning. It allows teams to see imbalances instantly and correct them on the spot. Instead of finding out months later that a diffuser was misaligned, we can fix it before the first experiment ever begins.

Closing Reflection

When you step into a laboratory, listen. The hum of air through ducts, the soft whisper of flow across the sash, those sounds tell you if the lab is breathing properly. A well-tuned system feels calm and effortless. A bad one feels restless, chaotic.

The laboratory ventilation system is the heartbeat of containment. When design, operation, and awareness move in sync, the result is more than compliance, it's safety you can see, hear, and feel.

Chapter 5

User Practices and Safe Operation

"The best-engineered hood can fail in the hands of a careless user."

You can have the most advanced fume hood ever built, but if the person standing in front of it doesn't understand containment, it's just an expensive piece of furniture. I've seen hoods that could pass every engineering test on paper but fail the moment a user leaned too close or piled boxes inside. Containment starts, and ends, with user behavior. The user is part of the overall system.

When I teach laboratory safety, I remind people that the fume hood isn't just equipment. It's a relationship between human and machine. You bring awareness; the hood brings protection. When one side stops paying attention, the partnership breaks.

The Human Factor in Containment

Containment begins at the sash. Every movement, every gesture, every decision in front of a hood affects airflow. Move too fast, and you stir turbulence. Raise the sash too high, and you break the air barrier. Even a quick reach inside can cause an invisible wave of contaminants to roll toward you.

Visualization technology has made this painfully clear. The

first-time people see a Tri-Color beam roll outward as they move, they instantly understand why rules matter. You can't unsee it, and that's a good thing.

Story from the Field:
During a training session in India, I asked a researcher to demonstrate how she normally worked. As she reached in quickly to grab a flask, the green beam of the Tri-Color laser flashed outward like a signal flare. The entire class gasped. No lecture could have made that point better. From that day on, her technique changed forever.

Common User-Induced Failures
Over the years, I've cataloged the mistakes I see most often:

- Overloading the hood and using it for storage
- Putting large objects directly on the worksurface
- Blocking the baffles
- Working too close to the sash opening.
- Storing boxes or instruments inside the hood.

- Moving too quickly when reaching in or out.
- Raising the sash beyond the safe operating height.
- Leaving the sash open when not in use.

Each of these behaviors changes the airflow pattern. Some block it, some accelerate it, and others simply confuse it. The hood can't protect you if you don't let it.

Key Insight:
A fume hood can't overcome human error; it can only reveal it.

Safe Work Practices
Good containment habits are simple but powerful. When they become routine, they transform how a lab feels and functions.

A forklift driver uses a checklist list everyday before he operates it. Why would a hood operator not do the same?

Here's the checklist I teach in every workshop:

1. Verify airflow before use.
2. Keep the sash at or below the safe height.
3. Work at least six inches inside the hood.
4. Move hands and materials slowly and deliberately.
5. Keep airfoils and baffles clear.
6. Remove clutter and minimize storage.
7. Close the sash when finished.
8. Report alarms immediately.
9. Always wear proper PPE.
10. Treat the hood as your partner in safety, not your servant.

Visualization as a Teacher

There's a reason why visualization training has taken off worldwide. People believe what they see. When users watch airflow react to their movements, they stop memorizing rules and start understanding reasons. In one university, after a single Tri-Color session, the percentage of hoods left open after use dropped by 80%. That's not compliance, it's comprehension. No "shut the sash" program, just a clear understanding of what the hood does, and how their action impact the entire lab.

Visualization transforms training into experience. It's one thing to tell someone that turbulence causes leaks, it's another to let them watch it happen.

Key Insight:

Seeing containment failure once teaches more than a

hundred safety posters ever could. And once you see it, you can't unsee it.

The Psychology of Seeing Containment
Human beings are visual learners. We trust what we see, even when we don't fully understand the science. When a user sees airflow behave correctly, confidence grows. When they see it fail, awareness deepens. Visualization makes that learning permanent.

I've watched entire safety cultures shift in a single afternoon once people understood airflow as a living, moving thing. That's the power of seeing.

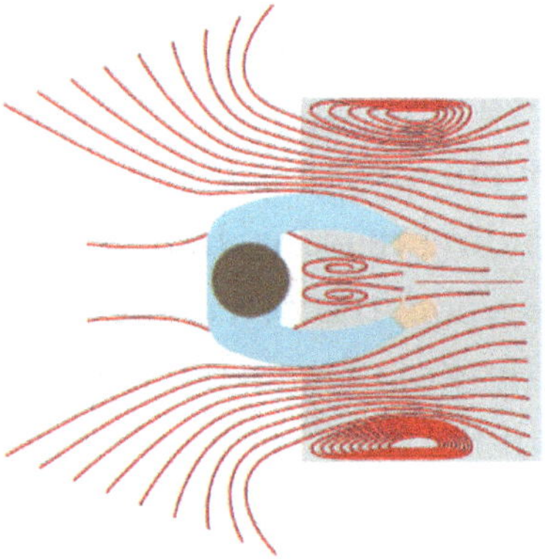

Integrating Training and Testing
More and more organizations are pairing annual performance testing with hands-on user demonstrations. During testing, technicians now take a few extra minutes to show users how airflow responds to their movements. It's containment education in real time.

I encourage every lab to make visualization part of its onboarding. Show new researchers what happens when the sash is raised too high or a door is left open. These small demonstrations build lifelong habits.

When I Certify Fume Hood Testers, I tell them that 50% of their job is to educate the users.

Key Insight:
Every containment test is an opportunity to train, not just certify.

Closing Reflection
Engineering controls protect people only when people respect them. The fume hood is one of the most powerful tools in the lab, but its strength lies in understanding. When users move with awareness, the system becomes flawless.

Containment isn't a mechanical achievement, it's a human one. And when we teach people to see air the way engineers do, safety becomes second nature.

Chapter 6

Troubleshooting and Diagnostics

"When containment fails, the air will tell you why, if you know how to look."

Every fume hood tells a story, and sometimes that story includes failure. But failure, if you listen carefully, is just information waiting to be understood. Troubleshooting containment isn't about guessing, it's about learning to read the air.

I've been called into hundreds of labs over the years to solve 'mystery leaks.' In nearly every case, the problem wasn't the hood itself, it was something in or around it. An HVAC change, a blocked baffle, or a nearby door left ajar. The air always gives you clues. You just have to know where to look.

Key Insight:
Containment failures rarely happen suddenly. They drift quietly into existence, and visualization is how we see them coming.

Data versus Seeing:

We saw with visualization there were occasional loss of containment over a 5-minute period, so we set up an anemometer 36" in front of the hood and recorded air velocity for 5 minutes, this is what we saw:

Face Velocity Traverse Variation

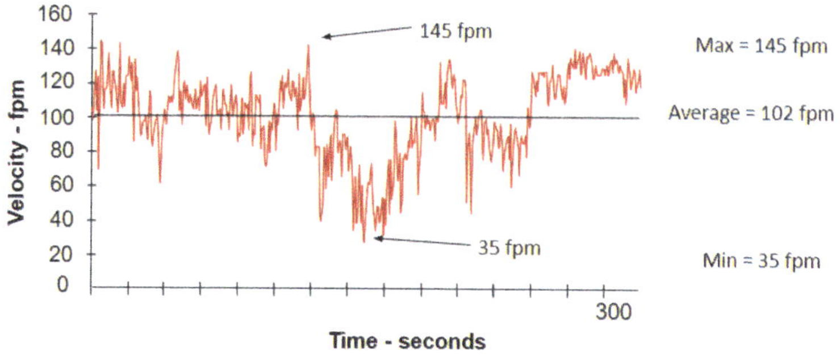

You can see that the average was 102 fpm and the monitor was hovering around 100 fpm. But you can see the minimum was 35 fpm. You can guess where we were seeing the green flashes of loss of containment. Yet an operator would look at Airflow Monitor and see numbers around 100fpm and assume the hood was working fine. These small changes are the air's way of whispering that balance has been lost. The most dangerous words I ever hear in a lab are, "It seems to be working fine".

The Diagnostic Mindset

Troubleshooting a fume hood is like detective work. You start with symptoms, form a hypothesis, and gather evidence. But unlike most mysteries, the suspect is often invisible. Visualization tools make that suspect visible.

Here's the mindset I teach technicians:

1. Assume the hood is innocent until proven guilty.
2. Look for environmental changes first.

3. Observe airflow before measuring it.
4. Always let visualization lead the investigation.
5. Look for instability in supply and exhaust air

In nearly every investigation I've led, visualization has solved the mystery in minutes. The lasers don't lie.

Story from the Field:
I was once called to a research facility where six hoods suddenly started leaking. Engineers suspected worn fans. We brought in the Tri-Color system, and within ten minutes, the culprit appeared, a new supply diffuser blowing directly at the hood faces. We changed the register and the containment stabilized instantly. The hoods weren't broken, The building was out of tune.

Tools and Techniques for Diagnosis
A good troubleshooter combines intuition with data. Here are the tools I rely on most:

- Camera: Usually on a phone to capture photos and videos
- Anemometer: Measures air velocity.
- Hazer Pro: Zero Velocity haze generator.

- Handheld Laser: Highlights the haze.

- Pressure gauge: Checks room and duct static pressure.
- Tri-Color system: Shows airflow behavior instantly.
- Q200 and SF6: Provides quantitative verification.
- Fume Hood Performance Tracker: Stores test history and patterns over time.

Each tool tells part of the story. But when you put them together, you see the whole picture.

Key Insight:

Data tells you something changed; visualization tells you where it changed.

Common Causes of Poor Performance

Over half of containment failures are caused by environmental or user-related issues, not mechanical defects. The most frequent culprits include:

- System imbalance after HVAC modifications.
- Cross-drafts from diffusers or open doors.
- Blocked or dirty baffles.
- Damper misalignment or fan wear.
- Excessive equipment inside the hood.
- Poor user technique.

Visualization can confirm each of these visually, often in seconds. Once you see the air behaving badly, fixing the problem becomes obvious.

Using Visualization for Diagnosis

Laser-based visualization is the fastest way to uncover the truth about airflow. The Tri-Color system, in particular, lets you interpret behavior at a glance:

- Blue beam shows a cross section of the vortex

- Red beams highlight airflow against the inside of the sash

- Green beams reveal containment loss.

Story from the Field:
At a university lab, containment was erratic, sometimes perfect, sometimes disastrous. Tri-Color revealed a hidden vortex forming each time the air conditioning cycled. Adjustments to the supply VAV fixed the problem.

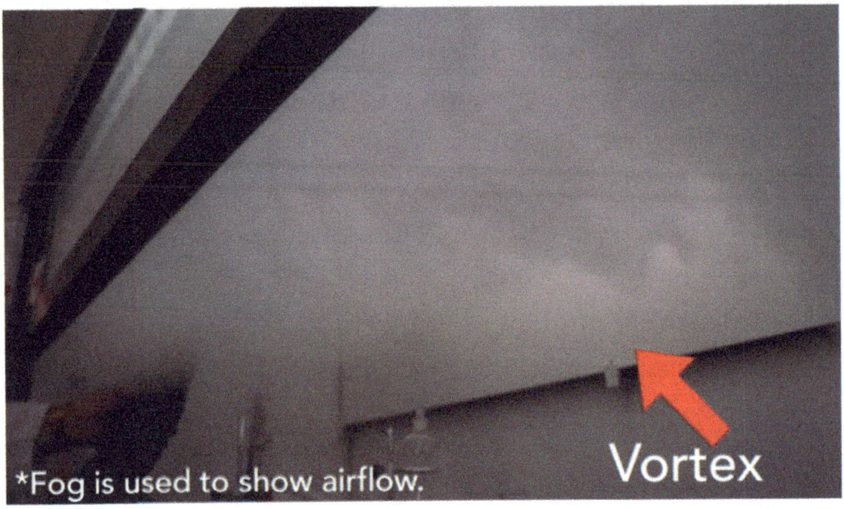

The need for Record Keeping
As a troubleshooter, finding good records was always a challenge. Without good history, it is incredibly difficult to understand and identify the problem. Fume Hood Certified developed a cloud-based solution called "The Fume Hood Performance Tracker" it is a digital repository for each individual hood.

The Digital Health Record for Hoods
In the past, test data lived in binders or faded stickers. Now, with tools like the Fume Hood Performance Tracker, every

hood can have a digital health record. Technicians scan a QR code, see the entire history of that hood, and instantly understand its performance trends.

We recently tested hoods at a major university and many of them were made by a company that ceased production in 1992. These hoods were at least 30 years old. Chances are no one who was around when the hoods installed were still here today or even remembered any of the history. You can believe that all that tribal knowledge and history are gone. Wouldn't it be great if it were all in a system like the Fume Hood Performance Tracker.

When troubleshooting, this context is invaluable. It reveals whether a problem is new or part of a recurring pattern. It's the difference between fixing a symptom and curing a cause.

Key Insight:
Every hood has a history. The best diagnostics begin by reading it.

Step-by-Step Troubleshooting Guide

1. Verify hood setup and sash position.
2. Check room conditions and nearby airflow sources.
3. Measure face velocity and compare to previous results.
4. Conduct visualization testing to identify turbulence or rollback.
5. Inspect mechanical components for wear or blockage.

6. Review digital history for recurring patterns.
7. Confirm containment restoration with final visualization.

Closing Reflection

Troubleshooting isn't about blame, it's about discovery. When something fails, the air is always trying to tell you why. Visualization lets you hear that story in color. Each containment breach becomes a lesson, not a loss.

When we listen to the air, we don't just fix problems, we prevent them from happening again.

Chapter 7
Energy Efficiency and Sustainability

"Safety and sustainability aren't enemies; they're partners in design."

For years, safety and energy efficiency were treated as opposing goals. You could have one or the other, but not both. If you lowered airflow to save energy, you risked loss of containment. If you increased exhaust to ensure safety, you burned energy. That false choice defined laboratory design for decades.

Today, that mindset is gone. We now understand that safety and sustainability are not competing, they're complementary. The key is validation. You can safely reduce airflow only when you can prove containment, and visualization has made that possible.

Key Insight:
Efficiency doesn't come from moving less air, it comes from moving the right air, in the right way, at the right time.

The Energy Burden of Fume Hoods
A single fume hood can exhaust 1,000 cubic feet of conditioned air per minute. Multiply that by hundreds of hoods in a large facility, and you quickly see why laboratories are among the most energy-intensive environments on earth.

Early attempts to reduce flow often failed because engineers treated airflow like a simple volume equation. But containment isn't about volume, it's about behavior. You can't just turn down the fan and hope for the best. You have to prove that containment still holds.

The Evolution of Low-Flow Design

The breakthrough came with low-flow hood design. Using Computational Fluid Dynamics and visualization, engineers discovered that stable containment could be achieved at lower velocities, if the hood geometry, baffles, and airfoils were optimized.

At first, this idea met resistance. People didn't believe that 60 fpm could be as safe as 100. Then the data came in. Facility after facility confirmed containment visually and saw dramatic energy savings.

Story from the Field:

A major research campus decided to cut hood velocities to 60 fpm across 30 laboratories. Before making the change permanent, they used Tri-Color visualization to confirm containment. The results were undeniable. Containment was perfect. Over two years, energy use dropped by 28%, and containment was perfect.

Variable Air Volume and Demand-Based Ventilation

The smartest energy savings come from flexibility. Variable Air Volume (VAV) systems adjust airflow in real time, responding to sash position, occupancy, and temperature.

When the sash is closed, the system throttles down airflow, saving energy. When opened, it instantly increases flow to help maintain containment.

Modern demand-based systems go a step further. They integrate occupancy sensors and predictive algorithms to fine-tune ventilation dynamically. It's a living, breathing balance between efficiency and safety.

Key Insight:
The greenest laboratories are the ones that know exactly when, and how much, to breathe.

Risk-Based Energy Design
ASHRAE's Laboratory Ventilation Design Levels (LVDL) framework brought risk-based thinking to ventilation. It allowed designers to tailor airflow intensity to actual hazard levels instead of arbitrary air change rates. This was a turning point for sustainability.

By aligning containment with risk, laboratories no longer had to over-ventilate low-hazard spaces. The result: lower carbon footprints, lower costs, and safer working conditions.

Visualization as an Energy Validation Tool
Visualization has become the bridge between sustainability and safety. It lets engineers prove that reduced airflow still protects users. I've watched facilities hold live demonstrations where the Tri-Color system showed

containment at 60 fpm in real time. You could feel the skepticism turn into confidence.

LVDL	Quantities of hazardous materials	Potential for Airborne Generation	Hazard Severity	Control Measures	Construction and Operating Costs	Types of laboratories / application
0	Negligible	Negligible	Negligible	Dilution Ventilation	Low	Computer and Instrumentation laboratories, secondary school teaching laboratories
1	Small	Negligible	Low	Dilution ventilation and local exhaust ventilation	Moderate	Secondary school teaching laboratories, shop areas with point sources of hazardous chemicals, biology laboratories without hazardous chemicals
2	Moderate	Low	Low to Moderate	Pull push dilution ventilation, local exhaust ventilation, fume hoods	High	Academic teaching laboratories, undergraduate research laboratories
3	Large	High	High	Pull push dilution ventilation, local exhaust ventilation, fume hoods, enclosure devices	Very High	Inorganic / organic synthesis laboratories, typical research laboratories
4	Very large	Very High	Extreme	Pull push dilution ventilation, local exhaust ventilation, fume hoods, enclosure devices	Extreme	Inorganic / organic synthesis laboratories, special high-hazard research laboratories, polymer synthesis laboratory, chemical development laboratories

One laboratory manager told me, 'We used to guess where the safety margin ended. Now we can see it.' That's the difference between belief and proof.

Story from the Field:
At a biotech company in California, visualization validated a 35% reduction in exhaust rates. They not only saved energy but also reduced noise and improved comfort. Seeing containment live on screen changed how everyone in the room thought about sustainability.

The Role of Smart Monitoring and Data Analytics
Continuous monitoring closes the loop. Sensors now record

airflow, pressure, and sash position continuously. These data streams feed into platforms like the Fume Hood Performance Tracker, which provides a health record for every hood.
With analytics, patterns emerged: sashes left open unnecessarily, ventilation rates exceeding design needs, or gradual drift in system balance. Facilities can now optimize ventilation based on real behavior, not assumptions.

Key Insight:
You can't manage what you can't measure, and now we can measure many aspects of performance.

Balancing Safety and Sustainability
The real challenge is cultural, not technical. We must replace the belief that more air equals more safety with the knowledge that smarter air equals better safety.
Visualization helps bridge that understanding. It gives engineers and safety officers the confidence to lower airflow without lowering protection.

Sustainability isn't a compromise, it's the logical outcome of understanding airflow.

Closing Reflection
When I look back at how far we've come, I realize that sustainability was never about cutting corners, it was about seeing clearly. Once we visualize containment, the old trade-off between safety and efficiency disappeared.

The best laboratories of the future won't choose between being green and being safe, they will achieve both. Because the moment you can see the air, you can finally move it wisely.

Chapter 8
Global Standards and Compliance
"Containment has no borders; safety speaks a universal language."

When I first began working with fume hood standards, many countries had their own playbook. The U.S. had ASHRAE 110, Europe had EN 14175. Each was good in its own way, but together, they created a patchwork of interpretations. For global companies and researchers, it was chaos. A hood that 'passed' in one country might fail in another.

Over the last five years, that's begun to change. We're witnessing a quiet revolution in harmonization. Engineers, safety professionals, and manufacturers across continents are starting to speak the same language: the language of containment.

Key Insight:
Containment standards may differ in method, but their purpose is the same, proving protection.

The Need for Global Harmonization

In a world where research collaborations span oceans and products cross borders daily, inconsistency in safety standards is more than a nuisance, it's a liability. Laboratories in multinational organizations need assurance that a hood designed in Germany will perform just as safely in Shanghai or Phoenix.

That realization has driven an unprecedented level of cooperation among international standards bodies. Committees from ASHRAE, SEFA, EN, ISO, and GB are now working toward common testing language and unified containment metrics.

The Major International Standards

Each standard has its strengths, and together, they form the backbone of global containment.

- ASHRAE 110 (United States): Focuses on quantitative testing using tracer gas and mannequin setups. It defines face velocity, smoke visualization, and containment verification.

- ANSI Z9.5 (United States): Is a more holistic view of Laboratory Ventilation and recommends some acceptance levels for ASHRAE 110.

- Tri-Color Dynamic Containment: Takes ASHRAE 110 and builds on its recommendation to do dynamic containment testing. Using this protocol, you can produce an ASHRAE 110 report.

- SEFA 1: Expands on ASHRAE by addressing construction and performance criteria, providing manufacturers with quality and durability guidelines.

- EN 14175 (Europe): Introduces dynamic containment testing, including moving mannequin and cross-draft simulations. It reflects real-world use conditions.

Key Insight:
The future isn't about choosing one standard, it's about merging their best ideas into a unified, global definition of containment.

Visualization: The Universal Language
If there's one thing that transcends borders, it's visualization. Whether you're in Beijing or Berlin, when you see a containment breach light up in green under a Tri-Color beam, you understand it instantly. Visualization has become the common language of safety.

During international Tri-Color demonstrations, I've watched audiences who didn't share a word of English nod in perfect understanding as airflow patterns came alive before their eyes. You don't need translation to see containment failures.

Regional Innovations and Adoption
Different regions have taken the lead in different aspects of containment advancement:
• United States: Integration of ASHRAE 110 with visualization-based commissioning.
• Europe: Refinement of dynamic containment testing using robotic mannequins and automated sash movement.
• China: Rapid adoption of visualization as part of the GB standard, driven by companies like Sunway Technologies.
• India: Growing use of risk-based design frameworks and training programs inspired by SEFA and ASHRAE.

Story from the Field:
In 2024, I stood in a test lab in Shanghai watching a Tri-Color system illuminate a EN-14175 compliant hood. The engineers were ecstatic. For the first time, they could demonstrate containment visually to clients. They didn't need to explain it,they could show it. That's how technology builds trust.

Bilingual Documentation and Collaboration
Global partnerships now demand clear communication. Test reports, compliance labels, and training materials are increasingly bilingual or multilingual. It's not just about translation, it's about transparency. When everyone can read and verify data, safety improves.

Working with International partners have shown how collaboration can drive innovation while maintaining compliance.

Key Insight:
Transparency builds trust. Trust builds safety.

Toward a Unified Global Standard
We're moving toward an era where containment will be verified the same way everywhere, through a combination of quantitative data and visual confirmation. Visualization is the great equalizer. It removes interpretation, making airflow behavior visible and universal.

Standards committees are already discussing hybrid approaches that merge ASHRAE's precision with EN's dynamic realism. The result will be a standard that speaks the same truth in every lab: protection proven by performance. You may see a worldwide standard come from China. They have no legacy to protect and few of the standards have been codified.

Validation of Design Intent:
We are starting to see specifications that that require specific testing to be performed for acceptance. It's no longer about testing in isolation; it's about integrating design intent with performance validation.

This same spirit of cooperation is spreading worldwide, proving that containment truly has no borders.

Closing Reflection

No matter where you go, a hood's purpose never changes, to protect the person standing in front of it. Whether it's tested under ASHRAE, EN, or some other criteria, the goal is the same: containment that can be trusted.

When I look at how far we've come in global collaboration, I see more than technical progress. I see a shared understanding, a universal respect for the invisible science that keeps people safe. Containment, after all, is a human value as much as an engineering one.

Chapter 9

Case Studies and Lessons Learned

"Every accident tells a story. The question is, are we listening?"

Every serious laboratory accident leaves behind two legacies: the event itself and the lessons we either learn or ignore. I've spent much of my career studying both. The stories that follow aren't just history, they're the reason we test, train, and teach.

Key Insight:
Every accident is a teacher. The only tragedy greater than the accident itself is failing to learn from it.

The Sheri Sangji Case - UCLA, 2008
In December 2008, 23-year-old research assistant Sheri Sangji was working with a pyrophoric chemical at UCLA when a fire broke out. She wasn't wearing a lab coat. The chemical ignited on contact with air, and within seconds, she suffered burns over nearly half her body. She died 18 days later.

It was one of the most devastating, and transformative, events in modern laboratory safety. It exposed gaps not just in equipment, but in culture: training, supervision, and awareness. The hood was there. The exhaust was running. But the system failed her in every other way.

Story from the Field:
After the incident, I was invited to speak at a university that wanted to make sure 'this never happens here.' During my presentation, I said something that's still true today: 'Safety isn't about compliance. It's about caring enough to make sure the people you work with go home at night.' That's what containment, and training, are really about.

The St. Andrews University Fire - Scotland, 2017
Nine years later, history nearly repeated itself in Scotland. A postdoctoral researcher working with flammable solvents experienced a fire that spread rapidly through a fume hood. The hood itself was not at fault; the fire originated from improper technique and inadequate supervision. The researchers survived, but the lab did not.

When investigators reviewed the footage, they saw what so many of us already know: the hood was never used properly. The sash was too high, the setup was crowded, and airflow couldn't protect the user from the ignition.

Containment is a partnership; it can't protect you if you ignore it.

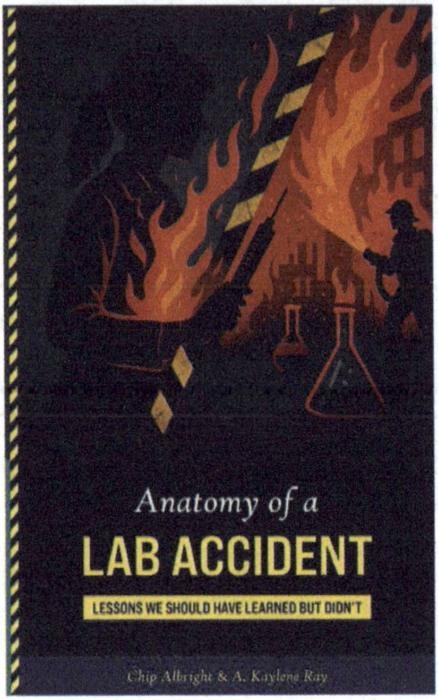

Story from the Field:
At a private pharmaceutical facility in the Midwest, technicians began noticing headaches after working long hours in the synthesis lab. No alarms had sounded. All hoods were certified and stickered as safe. Yet when we conducted a Tri-Color visualization test, the truth appeared, loss of containment was occurring from the bypass and being pulled down through the users breathing zone. When testing with SF6 no leakage was detected largely due to the fact that SF6 is 6 times heavier than air and little of it is in the upper part of the hood where the leak was occurring.

The certification said 'PASS.' The visualization said 'DANGER.' Within two days, the airflow was corrected, the symptoms disappeared, and an entire team gained a new understanding of containment.
When I asked one of the technicians what he learned, he said, 'I'll never trust a sticker again.' That's not cynicism, that's wisdom earned through experience.

Lessons from Success
Not every story is about failure. Some are about triumph. At John Brown University, after integrating visualization into their training program, the entire lab culture changed. Students began closing sashes instinctively. They adjusted their posture and slowed their movements. Accidents dropped to zero.

Visualization made containment visible, and once they could see it, they owned it.

Key Insight:
Containment is not just a technical goal, it's a behavioral habit.

Patterns We Can't Ignore
Across every case study, failures and successes alike, three truths emerge:
1. Containment must be verified, not assumed.
2. Training must be experiential, not theoretical.
3. Safety culture must be proactive, not reactive.

When I visit labs today, I don't ask to see their certification stickers. I ask to see their containment tests, their training logs, and their people in action. Because paper proves compliance; behavior proves understanding.

Story from the Field:
In one research lab, I asked a young chemist what she does before starting work. She smiled and said, "I have a checklist and a plan in case something goes wrong." That's when I knew we were finally getting somewhere.

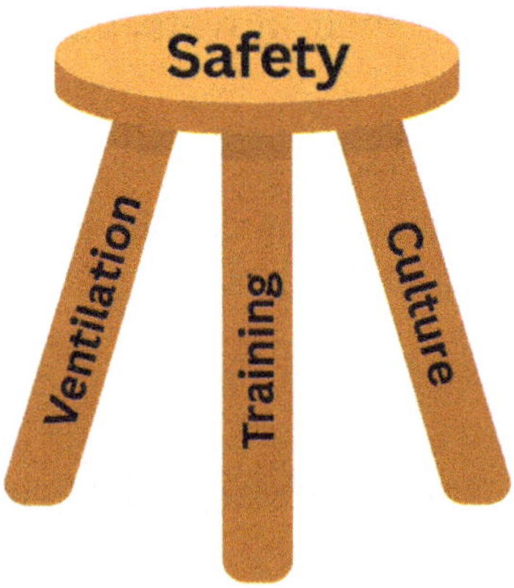

The Power of Remembering:
Every story in this chapter represents a life changed, or a life lost. Remembering them isn't just an act of respect; it's an act of prevention. When we forget, we repeat. When we remember, we evolve.

I often say that airflow tells the truth. But the truth is only useful if we're willing to see it, and act on it.

Closing Reflection

In the end, every accident is a teacher, and every safe day is a quiet success that often goes unnoticed. The laboratories that learn from the past are the ones building the safest future. Visualization ensures those lessons stay alive, because once you've seen containment fail, you'll never take it for granted again.

Chapter 10

Maintenance and Periodic Testing

"Safety isn't something you do once a year, it's something you prove every day."

For most of my career, maintenance and testing were treated as annual rituals. Once a year, a technician would show up, take a few velocity readings, put a sticker on the hood, and move on. It was the illusion of safety, an act of faith wrapped in a certification label.

But faith doesn't stop a leak. Airflow does.

Over the past five years, that mindset has changed dramatically. Laboratories have begun to realize that safety is not an event, it's a process. Maintenance is no longer about checking a box; it's about maintaining proof. The best facilities today treat testing as an ongoing dialogue between the hood, the ventilation system, and the user.

Key Insight:
Containment isn't verified once, it's verified continuously.

The End of Sticker Culture
For decades, the annual certification sticker was the symbol of safety. It told users, 'You're good for another year.' But that sticker didn't know if the HVAC system was rebalanced last month, or if the hood was overloaded this morning. It couldn't warn you when the exhaust fan started to slow.

Stickers gave comfort but not protection. The new standard is proof, real-time, data-driven, visible proof.

Continuous Validation and Predictive Maintenance
Modern fume hoods can now monitor their own containment performance. Sensors track airflow, sash position, and face velocity, feeding data to digital dashboards. If performance drifts, the system alerts maintenance staff automatically. This is predictive containment management.

Imagine a hood that tells you it needs attention before a user ever notices a problem. That's not fantasy, it's happening right now in the smartest laboratories in the world.

Story from the Field:
At one biotechnology lab, a smart hood system alerted staff to a 10% drop in exhaust pressure. They discovered a damper partially blocked by debris. Had they waited for the next annual certification, they would have operated unsafely for months. Instead, the issue was fixed within hours. The system didn't just monitor, it protected.

QR Codes and Digital Health Records

Every hood should have a history, just like a patient has a medical history. When you scan a QR code on a hood, you should be able to see every test result, every airflow trend, and every maintenance note. This digital transparency changes how users think about safety. They no longer rely on assumptions; they rely on evidence.

We see this a being similar to the maintenance log in a commercial aircraft. The primary reason they exist, is because the aircraft will be in service for many years, and thousands of pilots will fly it. When a new pilot gets on the plane, he first looks at the log to see if there are any issues. This assures that the plane is safe to fly.

Systems like the Fume Hood Performance Tracker have made this simple. Each hood becomes has its own record, a living document that evolves with every test or event.

Key Insight:
Transparency builds accountability. When you can see the data, you own the safety.

The Role of Visualization in Re-Certification
Annual testing has evolved beyond numbers. Visualization is now the heart of re-certification. Before a hood receives its new validation label, it must prove containment visually. This has turned testing from a procedural task into an educational event.

I've seen users crowd around during re-certification, watching airflow beams dance inside the hood. For the first time, they understood what the technician was measuring. They saw how their actions could affect containment. That awareness is priceless.

Story from the Field:
During a hospital lab re-certification, a young technician noticed air rolling out of a hood as the tester ran the visualization. The cause? A stack of bottles blocking the rear baffle. Within minutes, the problem was fixed, and the lesson was learned. She told me later, 'I'll never store anything in the back again.'

Maintenance Scheduling and Risk-Based Intervals
Not all hoods are created equal. Some run 24/7 with volatile chemicals; others are used sparingly for benign tasks. Testing intervals should reflect risk, not routine.

Unlike with Biological Safety Cabinets, there are no codified standards that mandate annual testing. Risks should determine maintenance schedules. High-use or high-risk hoods may require quarterly verification, while low-risk teaching labs can safely extend intervals with proper monitoring.

Key Insight:
Maintenance frequency should be determined by risk exposure, not the calendar.

The Human Element in Maintenance
Even the smartest system can't replace human observation. A technician who fully understands the holistic nature of the entire system can prevent more accidents than a thousand sensors. Training testers and maintenance teams to 'see and feel the air' is just as important as equipping them with instruments.

Visualization, again, becomes the bridge. When maintenance personnel can see how their adjustments affect containment, their work becomes proactive, not reactive.

Story from the Field:
At a research hospital, a maintenance engineer told me, 'I used to fix hoods. Now I protect people.' That's the mindset shift this technology creates.

The Future: Automated Assurance
We're not far from fully automated containment assurance.

Self-validating hoods will continuously monitor their airflow, send performance data to the cloud, and trigger service calls automatically. AI will predict when there are problems.

This isn't about replacing people, it's about empowering them. The goal is a lab where no one ever has to ask, 'Is it safe?' because the hood has already answered that question.

Closing Reflection

Maintenance and testing are not chores, they're acts of stewardship. Every time we verify containment, we honor the trust others place in our work. Safety is not something you confirm once a year, it's something you prove every day.

When I walk into a lab and see everything in sync, I don't see technology, I see a promise kept.

Chapter 11

Common Misconceptions

"The most dangerous thing in any lab isn't a chemical; it's a false belief."

In every training session I've ever led, I can predict the questions before they're asked. They all come from the same place, a handful of beliefs that refuse to die. These misconceptions are the ghosts of an older era of laboratory safety, and they still haunt too many labs today.

Key Insight:
What we believe about airflow determines how safely we work. Beliefs must evolve as understanding does.

Misconception 1: 100 fpm Means Safe
This is the granddaddy of all myths, the idea that face velocity equals containment. For decades, people were taught that as long as the hood averaged 100 feet per minute, it was safe. That number became gospel. The problem? It's not true.

Containment depends on stability, not speed. A hood with turbulent air at 100 fpm can leak more than a calm, balanced hood at 60 fmp. Velocity tells you how fast air moves; containment tells you where it goes.

Story from the Field:
I once tested a university teaching hood showing 105 fpm across the face, perfect by the numbers. But when we ran a Tri-Color visualization, the green beam rolled right out of the sash. The air was moving fast, but not in the right direction. We lowered the flow, balanced the system, and containment improved. Faster wasn't safer, it was worse.

Key Insight:
Containment is not a race. The goal isn't fast air, it's stable air.

Misconception 2: The Sticker Means Safety
A sticker tells you one thing: the hood passed some test at one point in time. It doesn't tell you how it's performing right now. Yet many people see a fresh sticker and assume they're protected. That's the illusion of certification.

I've seen hoods proudly displaying current stickers that failed containment the very next day, because the building conditions changed.

Key Insight:
A sticker is a snapshot. Containment is a story.

Misconception 3: Monitors Guarantee Protection
Digital airflow monitors are great tools, but they can also create false confidence. They measure velocity, not containment. And the velocity they are measuring is not a

real face velocity. It is a sidewall velocity or valve pressure change and then calibrated to mimic the face velocity.

When the monitor says 'SAFE,' it means the airflow speed is within a programed range, but that doesn't mean containment is holding.

Visualization testing has proven this repeatedly. I've seen monitors saying the hood velocity was in a safe range while green haze is escaping through the sash. The data looked fine. The air disagreed.

Story from the Field:
A researcher in a pharmaceutical lab told me, 'My monitor said everything was fine.' I replied, 'It's the same logic as saying the speedometer looked fine while you drove off a cliff.' Instrumentation is valuable, but only when it's understood.

Key Insight:
A monitor measures one dimension of performance. Containment lives in three dimensions.

Misconception 4: Containment Failures Are Mechanical
Most failures I investigate have nothing to do with broken parts. They're about human behavior, sashes left open, cluttered interiors, blocked baffles, or cross-drafts from supply air. We like to blame machines because they can't talk back, but the truth is, people create most containment problems.

Story from the Field:
In one chemistry building, three hoods failed the ASHRAE 110 containment test. Engineers were ready to replace the fans until we walked into the lab and saw the real cause: the researchers had stacked boxes and instruments up to the sash. We cleared the clutter, retested, and every hood passed. Sometimes the fix isn't mechanical, it's behavioral.

Key Insight:
If you can't see the back wall of your hood, neither can the air.

Misconception 5: Visualization Is Optional
When I first introduced visualization testing, people called it a gimmick. They said tracer gas was the 'real science.' But over time, the evidence spoke for itself. Visualization doesn't replace data, it translates it. It shows airflow in a way that anyone, anywhere, can understand instantly.

In labs across India, China, Germany, Spain, Columbia, Canada, Mexico, the U.S. and several other countries, visualization has become the preferred language of containment. Once you see airflow roll out of a hood, you never forget what it means. Visualization turns safety from a rule into a revelation.

Key Insight:
Data tells you there's a problem. Visualization shows you where it is.

The Cost of Misconception

Misconceptions are expensive. They waste energy, reduce safety, and erode trust. But more than that, they keep us from seeing the truth that's right in front of us. Each false belief we let die is replaced by understanding, and understanding is what keeps people alive.

Closing Reflection

When I visit a lab and hear someone say, 'We've always done it this way,' I know there's work to do. Progress in laboratory safety doesn't come from repeating old habits, it comes from questioning them.

The air doesn't lie. It never did. We just needed the tools, and the humility, to finally see it.

Chapter 12

Training, Education, and Safety Culture

"You can't build a safety culture with posters; you build it with people who understand why."

Every great laboratory I've ever visited had one thing in common: people who understood why safety mattered. Not because someone told them to follow a rule, but because they had seen the consequences of not following one. Education isn't just the transfer of information, it's the passing on of perspective.

In the private pilot world, it is said "In an emergency, a pilot never rises to the occasion – they fall back to the level of their training and practice." A book full of safety procedures means nothing if everyone doesn't understand and practice them.

When I started teaching containment, I quickly realized that showing was more powerful than telling. The moment a person sees airflow behave, the way it curls, drifts, or rolls back, their understanding changes forever. That's how real safety culture begins: with awareness you can't unsee.

Key Insight:
Education isn't about memorizing rules; it's about changing how people see risk. If they truly understand the hazards, they can do things to minimize them.

The Problem with Traditional Training
For decades, safety training meant PowerPoint slides, quizzes, and a signature on a form. It checked the compliance box but did little to change behavior. Most people left those sessions remembering little more than the free coffee. Safety is somewhat a boring topic. If you change the focus to accident prevention more people will identify with it. We have focused on Accident Prevention in factories for years.

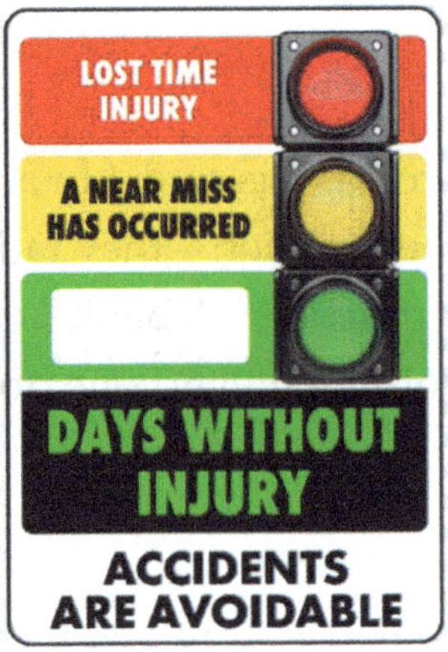

Safety culture failed not because people didn't care, but because they never experienced the 'why.' Visualization changed that. It turned theory into reality. When someone watches containment fail right in front of them, the lesson becomes permanent.

The Role of Visualization in Education

Tri-Color visualization is one of the most powerful teaching tools I've ever used. It bridges the gap between concept and comprehension. When air becomes visible, safety stops being abstract. You no longer have to explain why a sash should stay low, people can see what happens when it isn't.

I've seen rooms full of students gasp as green plumes of air roll toward them. I've seen professors stand silently, watching decades of assumptions unravel before their eyes. That's not training, it's transformation.

Story from the Field:

In Hyderabad, I once ran a workshop for a group of EHS professionals and engineers. During a demonstration, a senior engineer stood up and said, 'I've been testing hoods for 20 years and never truly understood what the air was doing until today.' That moment made the entire trip worth it.

Building a Culture of Curiosity

Safety isn't about fear, it's about curiosity. When people are encouraged to ask why, to explore, and to test, safety becomes a form of discovery. A strong safety culture is built not by enforcement, but by empowerment.

When you empower users to test containment themselves, to visualize airflow, and to understand how their actions change it, they become part of the process. That's how you build ownership.

Key Insight:
A true safety culture grows when users stop asking, 'What's required?' and start asking, 'What's right?'

From Compliance to Engagement
Compliance gets people to show up. Engagement gets them to care. The difference is emotional connection. Visualization creates that connection by making airflow behavior personal. When people see that their movements can cause containment loss, they begin to act differently, not because they were told to, but because they understand the stakes.

Story from the Field:
At one university, after adding visualization training to their orientation, the number of hoods left open overnight dropped by nearly 90%. Nothing else changed, just understanding. You can't regulate that kind of improvement. You can only inspire it.

Teaching the Next Generation
When I speak to the next generation of fume hood stakeholders, I always tell them that they're inheriting the next chapter of laboratory safety. The tools are smarter, the data richer, and the lessons deeper. But what matters most is still human behavior. Visualization lets them see the unseen forces that define their environment.

I often remind them: you are not just learning science, you're learning stewardship. The air that keeps you safe today will keep others safe tomorrow.

Key Insight:
A culture of safety isn't something you build once, it's something you teach every day.

Storytelling as a Teaching Tool
Stories connect where data cannot. When people hear about real accidents, like Sheri Sangji or the St. Andrews fire, they realize safety isn't theoretical. Every policy, every test, every sign on the wall has a story behind it. Stories make risk real.

When I tell these stories, I'm not trying to scare anyone. I'm trying to make them see the human side of safety. Because behind every rule is someone who paid the price for us to learn it.

Measuring Cultural Change
You can't manage what you don't measure, even when it comes to culture. The best organizations track training engagement, incident rates, and even hood's behavior over time. With digital tools like the Fume Hood Performance Tracker, safety culture becomes visible in data.

Over months, you can watch behavior trends improve, sashes staying closed more often, airflow events decreasing, fewer near-misses reported. That's what a living safety culture looks like: measurable, visible progress.

Key Insight:
Culture can't be faked, it has to be proven, just like containment.

Closing Reflection

Education is the single most powerful containment tool we have. The smartest hoods in the world mean nothing if the people using them don't understand airflow. But when knowledge, visualization, and care come together, something incredible happens, safety stops being a rulebook and becomes a shared belief.

You can't build a culture with posters or slogans. You build it one lesson, one story, and one moment of clarity at a time.

Chapter 13

The Future of Fume Hoods

"The next generation of fume hoods won't just protect, they'll think, learn, and teach."

If you had told me twenty years ago that a fume hood could think for itself, I'd have laughed. Today, I've seen it. The pace of change in laboratory containment has been nothing short of extraordinary. The future is no longer about stronger fans or faster air, it's about smarter systems that understand risk, adapt to change, and teach users in real time.

Key Insight:
Innovation in containment isn't about airflow, it's about awareness.

From Passive Protection to Intelligent Containment
For most of its history, the fume hood was a passive device. It did one thing, pull air. Now, with sensors, processors, and visualization built in, the hood has become an active participant in laboratory safety. It doesn't just respond to the environment; it interprets it.

Until recently hoods were slaves to the room. The room was the master and the hood obeyed. We are moving now to a peer-to-peer relationship where hoods and rooms talk to each other. But what is coming in the near future was unthinkable 10 years ago. As hood become smart and have a brain of their own, they will communicate with other hoods, these hives would look at how the building is performing in a holistic way. Then with the help of AI it will direct the building systems to optimize containment and reduce energy.

Imagine a hood that senses a sudden room pressure shift and adjusts exhaust in milliseconds. Imagine it alerts the user, changes lighting color to signal containment drift, and sends a notification to maintenance before performance degrades. That future is already here.

The Rise of the Smart Hood

Companies like Sunway Technologies have been leading the charge. Their smart hood systems integrate on demand containment validation through built-in visualization. When containment is verified, the hood glows green. When it's compromised, it glows red or amber. The user doesn't have to guess, the hood communicates directly.

But it doesn't stop there. These systems store performance data, predict maintenance needs, and even adjust airflow automatically based on activity levels. The hood has become both guardian and guide.

Story from the Field:
During a demonstration in Beijing, I watched as a Sunway Smart Hood shifted airflow patterns in real time after detecting a nearby door opening. The visualization beam flickered for a moment, then stabilized. Containment restored. The engineers in the room applauded. That was the first time I saw a hood 'think.'

Artificial Intelligence and Predictive Safety
The next evolution will be predictive containment. Using AI, smart hoods will analyze airflow trends, recognize patterns, and forecast risk before it happens. They'll learn from years of performance data and adapt to unique laboratory environments.

Think of it like preventive medicine for containment. Instead of waiting for failure, the system will anticipate it. It's no longer a question of 'What happened?' but 'What will happen if we don't act?'

Key Insight:
The future of safety lies not in reaction, but prediction.

The Role of Data Integration
Smart hoods are no longer isolated devices. They're part of a larger ecosystem of building management systems, digital safety portals, and energy optimization tools. Through platforms like the Fume Hood Performance Tracker, each hood's data feeds into a central network, providing facility-wide visibility in real time.

With integration comes accountability. Facility managers can see which hoods are performing well, which need attention, and which users might need retraining. It's the beginning of a new era, data-driven safety.

Visualization: The Interface of the Future
In the future, every hood will teach containment through visualization. Whether through laser projection, augmented reality overlays, or digital dashboards, users will see airflow behavior in real time. The hood itself will be the instructor.

Imagine walking up to a hood that greets you with a live airflow visualization, showing you exactly how your movement affects containment. That's not science fiction. It's already being prototyped.

Story from the Field:
At a technology lab in Singapore, engineers integrated an AR display into the sash glass. As the user moved their hands, airflow vectors appeared in real time. It turned training into interaction and safety into instinct.

Sustainable Design and Smart Efficiency
The smartest hoods will also be the greenest. They'll balance airflow precisely to minimize energy waste while maintaining absolute containment. They'll know when to rest and when to breathe deeply, adapting to occupancy and hazard levels.

Energy-efficient containment is not the enemy of safety, it's its evolution. Intelligent control turns sustainability into protection.

Key Insight:
Smart containment doesn't choose between safe and sustainable; it achieves both.

Global Collaboration and the Connected Laboratory
As laboratories become smarter, they're also becoming more connected. International research facilities will soon share containment data instantly across continents. Global databases of performance trends will help refine standards faster than ever before.

The days of isolated labs are ending. The future of containment is collaborative, transparent, and global.

Story from the Field:
In 2025, a joint testing initiative between Germany, India, and China used synchronized visualization across three labs to compare airflow behavior. The results not only improved each hood's design but also built bridges between engineering cultures.

The Human Future of Technology
Despite all the technology, one thing won't change: people. Even the smartest systems need curious, trained, and caring humans to interpret the data and act on it. AI can predict, but it can't care. It can warn, but it can't decide what safety means. That's still our job.

The smartest laboratories of the future will be those where people and machines learn together.

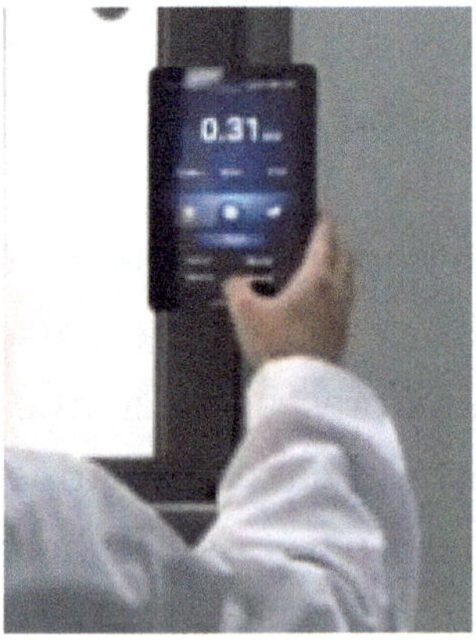

Key Insight:
Technology amplifies awareness, but awareness begins with us.

Closing Reflection

The next generation of fume hoods will not just protect, they will teach, adapt, and evolve. They'll be companions in safety, not just tools. The invisible will become visible, and the unseen will be understood.

As I often say, containment isn't about what the air does, it's about what we learn from it. The hoods of the future will make sure we never stop learning.

Chapter 14

Tri-Color: The De Facto Standard

"Once you've seen containment in color, you'll never trust invisible air again."

The first time I saw air come alive in color, I knew our industry would never be the same. It was during an early prototype demonstration, a narrow beam of light revealing what had always been invisible. When the airflow faltered, the color shifted, and the entire room fell silent. For the first time, we could see containment succeed, or fail, in real time.

That moment was the birth of Tri-Color.

Key Insight:
Visualization didn't just change testing, it changed understanding.

The Moment We Could See the Air
For more than fifty years, laboratory safety relied on numbers, gauges, and assumptions. We measured air velocity and hoped it meant containment. But airflow is like a river, it doesn't always follow the map. Tri-Color made that movement visible. Suddenly, the invisible enemy had form, color, and motion.

In one of my earliest workshops, a young engineer raised his hand after a demo and said, "I've been testing hoods for ten years, and this is the first time I've actually 'seen' what I've been measuring." That's when I knew Tri-Color wasn't just a testing tool, it was a teacher.

How Tri-Color Works

The Tri-Color system projects laser planes, red, green, and **blue**

- Blue shows a cross-section of the vortex
- Red reveals air movement inside the hood against the sash
- Green indicates loss of containment.

Unlike smoke, which is dense and can obscure visibility, Tri-Color uses Haze that is thinner and provides a continuous, dynamic view of airflow behavior. It's real-time visualization, safe for use during active lab operations, and environmentally friendly, no gases, no residue, no downtime.

Key Insight:

Tri-Color doesn't test the air, it teaches it to speak.

Visualization as Validation

For decades, tracer gas testing was considered the gold standard for containment verification. It provided quantitative results but lacked immediacy. Tri-Color bridged that gap. It gave engineers and users alike an instant understanding of performance.

When airflow rolled back, you could see it. When a diffuser disrupted the face, you could prove it. Testing became storytelling, each beam of light narrating the hood's truth.

Story from the Field:

At John Brown University, Tri-Color became part of their

testing protocol. Students didn't just read about containment, they saw it happen. Their lab culture shifted overnight. The data didn't change; the understanding did.

Adoption and Global Impact

What began as a testing and educational tool quickly evolved into an international method to test dynamic containment. In India, workshops drew hundreds eager to see containment made visible. In China, manufacturers like Sunway Technologies integrated Tri-Color directly into their smart hoods. In Europe, it became a supplemental validation method alongside EN14175 dynamic testing.

Today, more people in more countries use Tri-Color as a containment verification method. It's not a competitor to tracer gas, it's its complement, adding clarity where numbers alone can't speak.

Key Insight:
Visualization is the one language every safety professional understands.

Applications Beyond Testing
Tri-Color's strength lies in versatility. It's now used for:
• Performance Testing: Real-time airflow visualization during commissioning and certification.
• Training: Teaching airflow awareness to students and lab users.
• Diagnostics: Identifying turbulence, cross-drafts, and system imbalances.
• Design Validation: Supporting CFD results during hood development.
• Continuous Verification: Integrated into smart hoods for ongoing monitoring.

I've seen Tri-Color used in classrooms, factories, and cleanrooms. Everywhere it goes, it changes how people see safety, literally.

The Rental-to-Purchase Revolution
One of the smartest decisions we ever made was offering Tri-Color as a rental-to-purchase system. It allowed smaller institutions to access the technology without the upfront cost. Many rented it for a single demonstration, and never

gave it back. They saw what it could do and realized they couldn't operate without it.

Visualization sells itself. Once you see it, you believe it.

Key Insight:
When you make containment visible, you make it valuable.

Why It Became the **De Facto Standard**

The reason Tri-Color became the world's De Facto Standard containment validation method is simple: it works, it teaches, and it unites. Standards bodies reference visualization, but Tri-Color standardizes it. Manufacturers design around it. Users demand it.

It turned a once-abstract idea, dynamic containment, into something tangible and trustworthy. And perhaps most importantly, it gave safety professionals a way to communicate performance instantly to anyone, anywhere.

Closing Reflection
Tri-Color started as a question: what if we could 'see' the air? It became an answer, proof that seeing is believing. It has helped thousands of people understand containment, not as theory, but as reality.

When you turn on that laser and see the beams cut through the air, you realize something profound: safety isn't invisible anymore. It's alive, visible, and unforgettable.
Once you've seen containment in color, you'll never trust invisible air again.

Chapter 15

The Fume Hood Medical Portal

"Every fume hood has a life story; it's time we started keeping its medical record."

If you've ever worked in a hospital, you know the importance of a patient's medical chart. It tells you their history, the tests they've had, their medications, their recovery, and their risks. Now imagine if every fume hood in every lab had the same thing, a living, breathing record of its health and performance.

That's what the Fume Performance Tracker is all about.

Key Insight:
You can't manage what you don't track, and hood performance is no exception.

From Stickers to Systems
For decades, our only record of a fume hood's performance was a sticker in the corner of the sash. It told you when it was last certified and when it was due again. That was fine in a world where testing happened once a year. But in today's world of smart containment, it's like trying to diagnose a patient based on a note written twelve months ago.

The Fume Hood Performance Tracker changes that by creating a continuous, digital health record for every hood, its own medical chart. Every test, every airflow reading, every visualization result along with reference materials and maintenance records are stored and accessible.

The Anatomy of a Hood Record

Each hood's profile includes its make, model, location, date of installation, test results, maintenance notes, and even its visualization performance videos. Technicians can scan a QR code on the hood to see its full record instantly. And users can see a digital certificate.

HOOD
IDENTIFICATION
NUMBER

26-R24 Y44

SCAN ME

FHC FUME HOOD
CERTIFIED

Story from the Field:

At one research facility, the Tracker revealed that several hoods were gradually losing containment over months. A fan belt was wearing out, but instead of waiting for failure, maintenance replaced it proactively. The hood never failed containment, and the users never knew there had been a risk.

Turning Data into Prevention

Data by itself doesn't create safety, understanding does. The power of the Fume Hood Performance Tracker lies in its ability to translate thousands of data points into simple, actionable insights. It captures tribal knowledge and converts it into actionable data.

The system learns from patterns. If airflow trends start to deviate, it flags the issue long before containment is compromised. That's predictive safety in action.

Key Insight:

Every hood tells a story. Data, photos and video gives it a voice.

Integration with Smart Laboratories

The Fume Hood Performance Tracker isn't just for individual hoods, it's a network for entire facilities. It tracks performance across floors, buildings, and campuses. Facility managers can see the health of every hood.

The Power of Visibility

The beauty of this system is in its simplicity. It makes containment visible, not just during testing, but every day. Technicians no longer need to hunt through binders or spreadsheets. Users can instantly verify that their hood is safe before they start work.

This visibility builds trust. And trust builds culture.

Story from the Field:
One laboratory director told me, 'For the first time, I can tell our users that safety isn't a sticker, it's a system.' That's the shift this technology represents.

From Monitoring to Mentoring
What makes the Fume Hood Performance Tracker special is not just the data, it's the knowledge it shares. Over time, it will help identify which designs, manufacturers, and configurations perform best for you. It teaching you what works and what doesn't.

This digital repository can track anything that you think is important. It is like a baby book showing the hood's life.

Key Insight:
When data becomes shared knowledge, safety becomes shared progress.

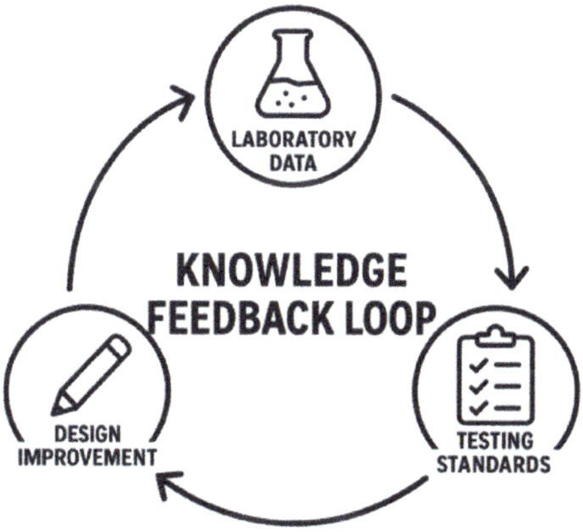

The Future: A Digital Containment Ecosystem
As more and more people use the system; it is continually updated based on what people tell us they need. The next step is integration with intelligent hoods and AI analytics.

This is how we move from reactive maintenance to proactive prevention, from paper stickers to predictive protection. As we move from analog to digital, massive amounts of data can be captured and customized to fit the needs of the organization.

Closing Reflection
Every hood deserves a record. Every record deserves to be read. When we start treating containment the way doctors treat health, through vigilance, diagnostics, and data, we stop reacting to failures and start preventing them.

The Fume Hood Performance Tracker is more than a database, it's a commitment to accountability. Because the air we trust with our lives deserves the same attention as the patients we protect.

Epilogue

Making Labs Safer, One Fume Hood at a Time

"Progress in containment begins with curiosity, and ends with someone deciding to make it safer."

When I look back on my journey through the world of laboratory containment, I realize it was never just about airflow or numbers. It was about people, students, scientists, engineers, and technicians, each standing in front of a fume hood trusting it to protect them. Every improvement we've made, every lesson we've learned, has been about honoring that trust.

We've come a long way from the days when safety was measured by a sticker. We now see containment as a living, breathing system, dynamic, measurable, and visual. We've learned that the air itself tells a story, and that story is written in color, data, and awareness.

Key Insight:
Containment isn't a standard to meet, it's a promise to keep.

The Journey of Change
Over these years, I've watched entire laboratories transform, not just through new technology, but through a shift in

mindset. The moment people 'see' containment, they understand it. The moment they understand it, they own it. From the first Tri-Color demonstration to the development of smart hoods and the Fume Hood Performance Tracker, the goal has always been the same: make the invisible visible, and make the complex understandable. That's how progress happens, not through mandates, but through meaning.

A Global Conversation

What started in a handful of labs has grown into a worldwide community of people who care deeply about safe containment. From Phoenix to Mumbai, from Beijing to Berlin, the conversation is the same: how can we make labs safer for the people who work in them? Each region brings its own ideas, its own energy, and its own commitment. Together, we're writing the next chapter of containment.

Key Insight:

Containment has no borders. Its purpose is universal, to protect those who pursue knowledge.

The Legacy of Laser Visualization

If I had to choose one contribution that changed everything, it would be laser visualization. It gave us the power to see the truth about airflow, to see not just whether containment worked, but 'how' and 'why' it worked. It turned data into dialogue, and testing into teaching.

That's the legacy I hope continues long after I'm gone: a

world where no one assumes safety, they prove it. Where every hood, every test, and every user interaction adds to a larger body of knowledge.

A Message to the Next Generation

To the next generation of laboratory professionals: never stop asking questions. Never stop challenging assumptions. And never stop looking for ways to make things safer, simpler, and smarter.

Technology will evolve, but the heart of safety will always be human. You are the guardians of that legacy.

Story from the Field:

At one of my final workshops, a young engineer approached me and said, 'I used to think fume hoods were just equipment. Now I see them as people's last line of defense.' That one sentence told me everything I needed to know about why this work matters.

The Promise We Keep

Safety is not a task we complete; it's a value we live. Every test we run, every standard we write, every hood we validate adds to that promise. The promise that everyone who walks into a laboratory will walk out safely at the end of the day.

That's why I end every presentation the same way: Making Labs Safer, One Fume Hood at a Time. It's not just a slogan, it's a mission.

Because in the end, containment isn't about air. It's about people.

Chip Albright
Phoenix, Arizona
2026

Free Fume Hood Stickers for Safety Training

If you would like some of these labels reach out to
experts@fumehoodcertified.com

Get Hoodie Gear

https://fumehoodtesterscatalog.com/collections/hoodie-gear

Take advantage of Fume Hood Certifier's extensive resources!

fumehood.academy
Information regarding our in-person workshops in Phoenix. A great event for any experience level.

fumehoodcertified.com
Learn about all that FHC has to offer and how we may able to help you

tricolorasertesting.com
Home of the Tri-Color Visualizer system. Learn how we utilize a modified ASHRAE110 large volume test to challenge fume hood's dynamic containment abilities.

fumehoodonlinecourse.com
Chip's book, "Laboratory Fume Hoods Explained" converted into a free online course

fumehoodperformance.com
The home for all your fume hood testing data for easy sharing, review and analysis. Works seamlessly with our Q200 and Face Velocity Profiler data capture software

fumehoodsafety.com
We host several introductory level videos here for students and employees that are new to fume hoods to get a understanding of safe user practices

fumehoodsoftware.com
Learn all about our Fume Hood Performance Tracker here, includes a walk-through video and a live demo.

fumehoodtesterscatalog.com
A one stop shop for everything you need to perform ASHRAE110 compliant fume hood testing, from smoke testing to Q200 factory calibrations, we have it all!

fumehoodtestingnetwork.com
Directory of industry professionals that utilize our Tri-Color testing system.

Safety is No Accident (Blog)
https://fumehoodtesterscatalog.com/blogs/lab-safety-digest

YouTube
www.youtube.com/@FumeHoodCertified

LinkedIn
www.linkedin.com/company/fume-hood-certified-llc

Facebook
www.facebook.com/fumehoodcertified

Laboratory Fume Hoods Explained (Free Online Course)
www.fumehoodonlinecourse.com

Bulk Sales

Help Spread the word.

We are offering large discounts for multiple copies. Reach out to us at experts@fumehoodcertified.com

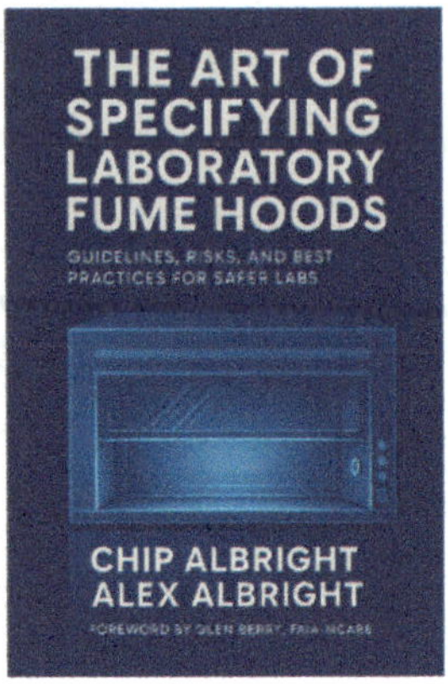

Do you have a Medical Portal?

Just like people use medical portals to monitor their health, laboratory fume hoods now have their own health tracker-**Fume Hood Performance Tracker**.

It's your digital portal to ensure every fume hood is operating safely and efficiently.

Fume Hood Certified is offering a **free trial** to help you get started. Discover how easy it is to track performance, flag issues before they become problems, and document safety like never before.

Don't wait for a safety incident to start monitoring your hoods-be proactive.

Email us today at **experts@fumehoodcertified.com** and start your free trial of the Fume Hood Performance Tracker.

HOOD
IDENTIFICATION
NUMBER

23-R24 Y44

SCAN ME

FHC FUME HOOD CERTIFIED

Your lab's safety deserves the same attention as your own health. Let us help you make that possible.

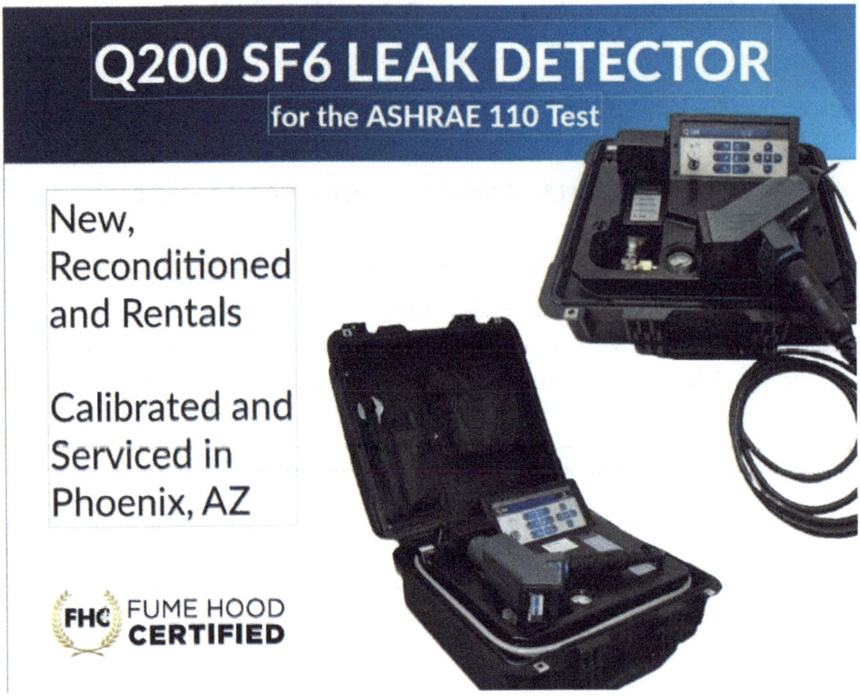
Q200 SF6 LEAK DETECTOR
for the ASHRAE 110 Test
New, Reconditioned and Rentals
Calibrated and Serviced in Phoenix, AZ
FHC FUME HOOD CERTIFIED

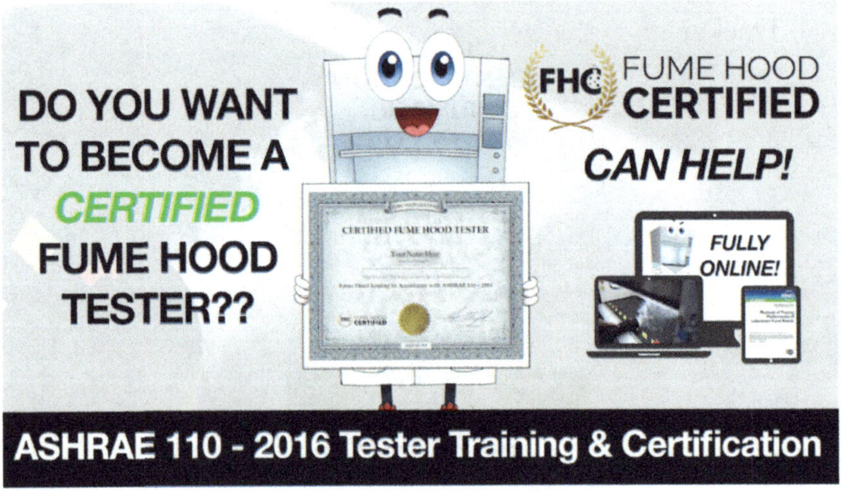
DO YOU WANT TO BECOME A CERTIFIED FUME HOOD TESTER??
FHC FUME HOOD CERTIFIED
CAN HELP!
FULLY ONLINE!
ASHRAE 110 - 2016 Tester Training & Certification

www.fumehood.academy

FHC THE CONTAINMENT CHRONICLES

www.fumehoodcertified.com/fhc-newsletter

OVER 100
PRODUCTS
FOR
FUME HOOD
TESTERS

www.fumehoodtesterscatalog.com

Made in United States
Orlando, FL
04 November 2026

90202254R00066